COMBAT AIRCRAFT

© Aladdin Books Ltd 1989

*First published in the
United States in 1990 by*
Gloucester Press
387 Park Avenue South
New York NY 10016

Design David West
Children's Book Design

Editorial Lionheart Books

Researcher Cecilia Weston-Baker

Illustrator Ian Moores

Printed in Belgium

All rights reserved
ISBN 0-531-17205-8

Library of Congress Cataloging
Card Number 89-81600

CONTENTS

Working parts	4
Different types	6
The jet engine	8
The wings	10
Take off	12
Flying the aircraft	14
Navigation and radar	16
Weapons	18
Attack	20
Defense	22
Dogfight	24
The future	26
History of combat aircraft	28
Glossary	30
Index	32

HOW · IT · WORKS

COMBAT AIRCRAFT

IAN GRAHAM

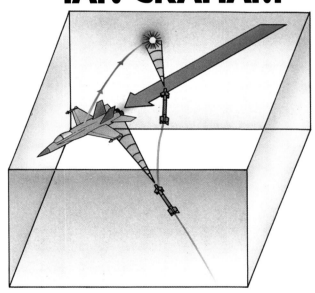

GLOUCESTER PRESS
New York · London · Toronto · Sydney

WORKING PARTS

A combat aircraft is a flying weapons machine. The McDonnell Douglas F-18 Hornet shown here is 18.6 yards long, 13 yards across its wing-tips and, with a full fuel and weapons load, weighs 21 tons. It is powered by two turbofan engines, each producing almost 16,500 lbs of pushing force, or thrust.

In flight, the aircraft is controlled by three major structures. Ailerons in the trailing edges of the wings hinge, one up and the other down, to make the craft roll to one side or the other. In the tailplane, elevators move up or down together to lower or raise the tail.

Rudders in the tail fins hinge to the left or right and push the tail in the opposite direction to help turn the craft.

Apart from these basic "control surfaces," which most aircraft have, combat aircraft have a number of extra

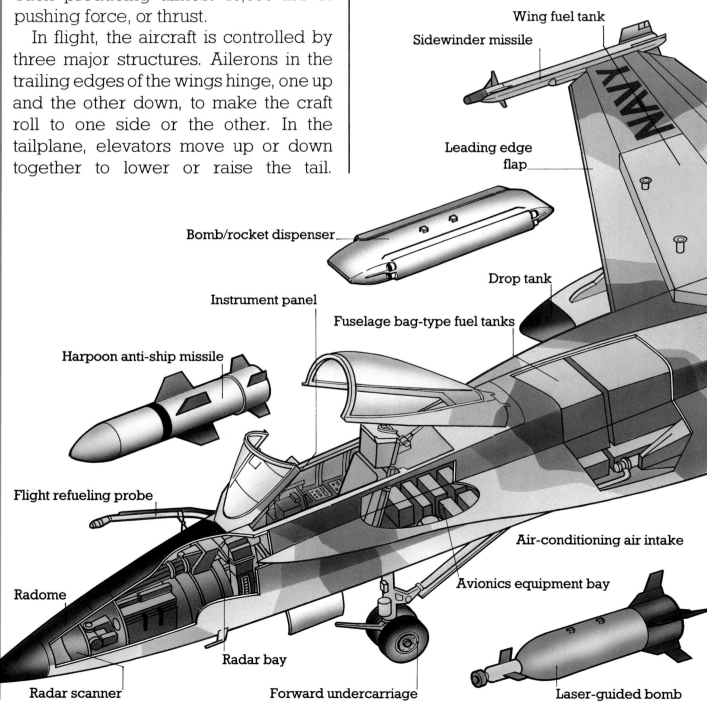

features. A panel in the tail section, called an airbrake, can be raised to slow the aircraft down. The F-18 has wings with both leading and trailing edge flaps to improve its performance at a range of speeds and in tight combat maneuvers. Attachment points called pylons enable bombs, rockets and missiles to be hung underneath the aircraft. Finally, the F-18 is packed with electronic systems including a sensitive radar scanner inside its nose-cone.

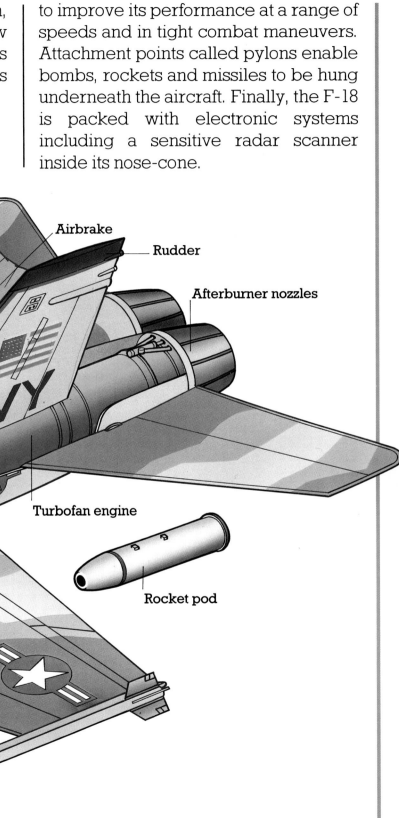

DIFFERENT TYPES

There is no such thing as a typical combat aircraft. There are several different types. Their shape, speed, the weapons they carry, and the height at which they normally fly, are all influenced by the various types of jobs they have to do.

Interceptor craft all specialize in attacking high-altitude bombers. They must be able to reach the bomber's altitude very quickly and fly to the bomber's position at high speed. They therefore need a high rate of climb and maximum speed, but they do not need to be very maneuverable.

Maneuverability is very important to Air-superiority fighters. These are designed for air combat against other fighters. Attack aircraft operate close to their targets on the ground. They require good maneuverability at low altitude and speed. Naval fighters have to operate from aircraft carriers. To do so, they must have good low-speed performance to enable them to take off and land on the short runway on the ship's deck.

As all warfare, or military, aircraft are so expensive, many are designed to do several jobs. The Fighter-bomber, for example, combines the functions of the fighter and the bomber. For one type of mission, an aircraft might carry a range of bombs. For another mission, the same aircraft might carry air-to-air missiles. Like the aircraft, the weapons – guns, missiles, rockets and bombs – are also specialized for different missions.

The Soviet MiG-29 high-altitude Interceptor.

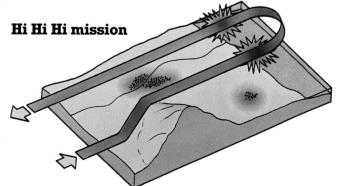

Hi Hi Hi mission

A typical Interceptor mission takes place at high (Hi) altitude throughout, because this is where the attacking aircraft will find its prey. The Interceptor's job is to defend territory against incoming bombers.

The Panavia Tornado multi-role aircraft.

Hi Lo Hi mission

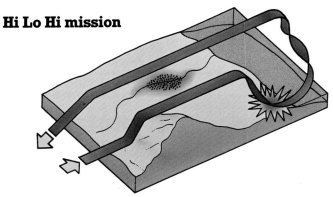

Air-superiority fighters usually attack low-flying aircraft from high altitude. After a successful attack, they return to high altitude. A typical Air-superiority mission therefore has a Hi Lo Hi profile.

The A-10 Thunderbolt Ground-attack aircraft.

Lo Lo Lo mission

Ground-attack aircraft fly Lo Lo Lo missions – often only yards above the land – not only because their targets are on the ground, but also to avoid the Air-superiority fighters that may be prowling at high altitude.

THE JET ENGINE

Modern combat aircraft are powered by one or more jet engines. In a jet engine, air is heated and so expands, creating thrust. With the gas turbine – the power unit that is commonly called a jet engine – cold air is sucked in at the front and then heated. The air expands and forces its way out fast through the engine's exhaust pipe. It is this "jet" of hot exhaust gases that gives the engine its popular name.

The first jet engines were designed and built by Sir Frank Whittle in Britain in 1930. They enabled aircraft to fly much faster than was possible with the piston engines used at the time. These were similar to car engines.

There are different types of jet engine. The most common is the turbofan. A large, many-bladed fan at the front of the engine draws in air. Only part of the air is sucked into the engine and heated; the rest just flows around it. This type of engine is quieter than the simplest type of jet engine, called a turbojet. A gas turbine can also be used to spin a propeller. In this case, the engine is known as a turboprop.

Air sucked into an engine by a turbofan passes through a compressor. This device "squeezes" the air, thereby increasing the air pressure inside the engine. The air is then forced into a combustion chamber. Fuel is sprayed into this and the air-fuel mixture is burned. The hot, expanding gases that are produced rush out of the engine through a

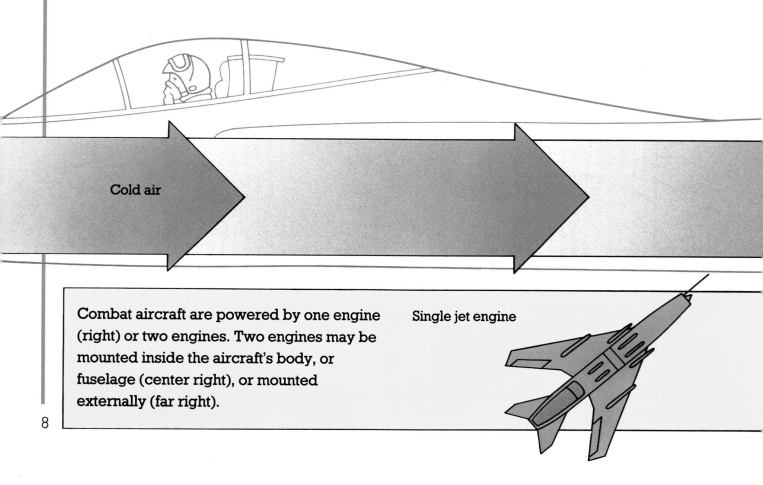

Cold air

Combat aircraft are powered by one engine (right) or two engines. Two engines may be mounted inside the aircraft's body, or fuselage (center right), or mounted externally (far right).

Single jet engine

Military aircraft use their afterburners in short bursts.

turbine, which drives the compressor. Engine power can be boosted by burning more fuel in the exhaust. This process is often called afterburning or reheat.

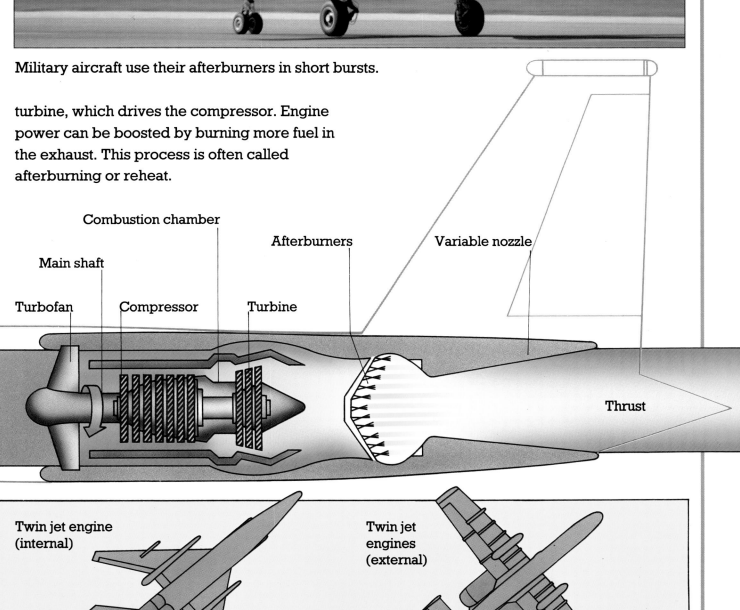

THE WINGS

All combat aircraft are heavier than air. In order to fly, they must create an upward force, called lift, to overcome their weight. This is done by the wings.

As an aircraft travels forward on the ground, its wings divide the air streaming past the craft into two flows. Some air is deflected over the top of each wing and the rest flows underneath it. The wing is shaped so that air has to travel farther over the curved top surface than the flatter lower surface. This makes the air above the wing flow faster than the air underneath it. This produces a difference in air pressure above and below the wing, and it is this pressure difference that produces the upward force of lift.

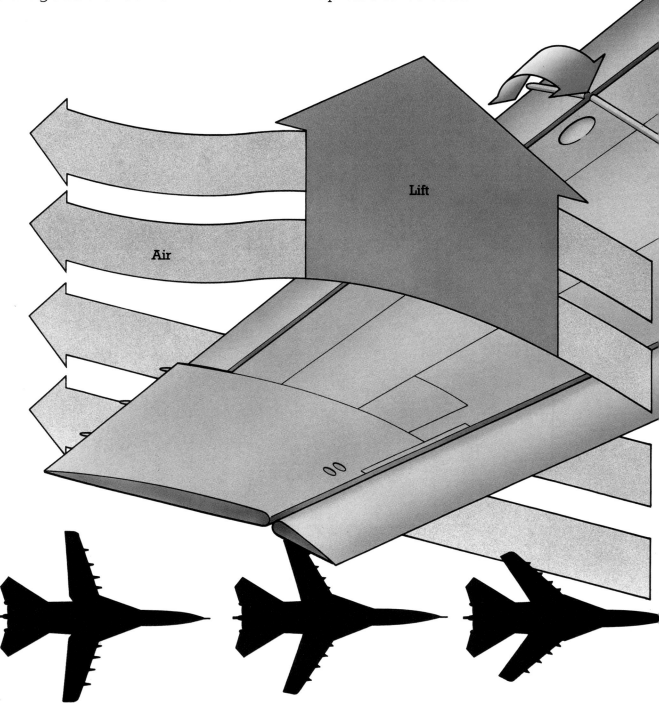

If an aircraft's speed increases, the difference in air pressure produced by the wings increases and so the amount of lift produced also increases. When the lift becomes greater than the aircraft's weight, the aircraft begins to rise into the air. Equally, as an aircraft slows, the amount of lift decreases and the craft begins to descend.

At very low air-speeds, the air flow around the wings may break up. If so, the wings lose all lift, which is called stalling. Movable extensions along the edge of the wings, called leading edge slats, help stop the air flow from breaking up and maintain lift.

In swing-wing aircraft, the wings are spread wide for slow flight and to carry heavy loads. With fully swept-back wings, drag is reduced and high-speed flight is possible through even the strongest of oncoming winds.

An aircraft's wings are shaped according to its speed and agility. The A-10 Thunderbolt's broad straight wings are best for slow flight and tight turns. Swept-back wings are better for high-speed flight. The delta is the most efficient shape for supersonic (faster than sound) flight. Small wings called canards on the aircraft's nose increase lift during take off and landing. Extending flaps along the trailing edge of wings helps to stop the aircraft losing lift.

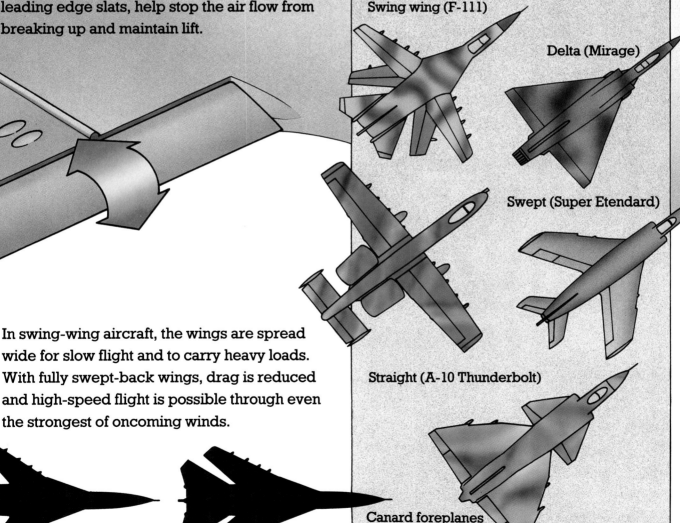

Swing wing (F-111)

Delta (Mirage)

Swept (Super Etendard)

Straight (A-10 Thunderbolt)

Canard foreplanes (Saab Viggen)

TAKE OFF

Most aircraft take off by accelerating along a runway until the lift produced by their wings is greater than their weight and they climb into the air.

Short take off and vertical landing (STOVL) aircraft like the Harrier "jump-jet" can use their engines to provide extra lift for take off and landing. The jet exhaust leaves the engines through four nozzles, two on each side of the aircraft. By rotating all four nozzles, the jet exhaust can be pointed in different directions.

With the nozzles pointing downward and to the rear, the engines provide thrust for accelerating the aircraft along a runway and also lift. The extra lift from the downward pointing nozzles enables the Harrier to take off from very short runways. If necessary it can take off vertically, but this uses up a great deal of fuel and so reduces the craft's range.

Harrier taking off vertically.

For vertical take off, the Harrier has its engine nozzles pointing downward. Engine thrust is increased until it overcomes the aircraft's weight. As the aircraft rises, the nozzles are slowly angled backward giving thrust for forward flight.

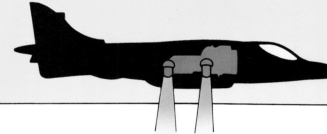

Some Harriers are specially designed to take off from and land on ships. These Sea Harriers normally operate from ships fitted with a ramp called a "ski jump." Other aircraft, such as the F-14 above, are launched from aircraft-carriers by steam catapult.

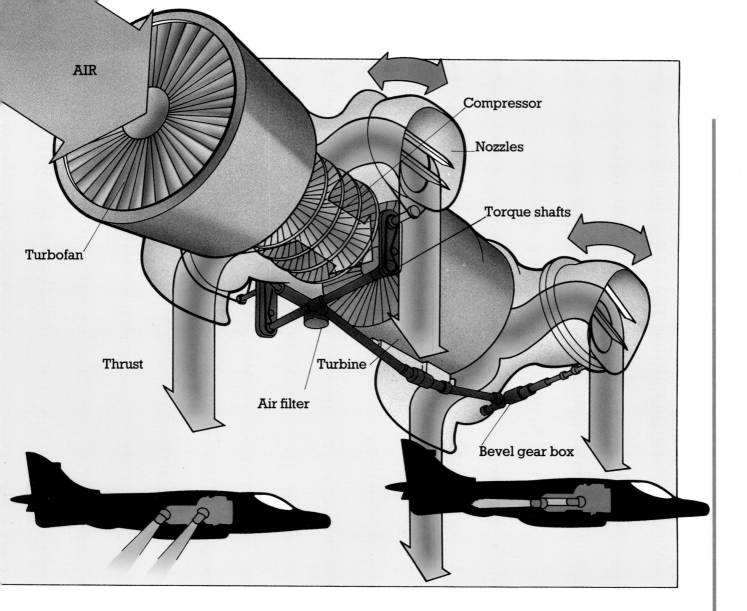

With its double-delta wings, these Saab Viggens can take off even from highways!

FLYING THE AIRCRAFT

If the pilot of a passenger aircraft were to let go of the controls, the plane would continue to fly straight and level. By contrast, a combat aircraft would fly all over the place because it is not designed to be stable naturally. This makes it more nimble in the air, but also more difficult to fly. Only computers work fast enough to make tiny adjustments to the control surfaces – perhaps 100 times every second – to keep the aircraft under control.

As the pilot of a combat aircraft moves the flight controls, their movements are turned into electrical signals. These pass along wires to computers, where they are interpreted and processed in order to operate the plane's control surfaces. The system is known as fly-by-wire. Other systems, like navigation and terrain following radar, also send data to the flight computers.

Important data is projected onto a glass screen in front of the pilot. All the data on this Head-Up Display (HUD) can be seen without having to look down into the cockpit.

If a combat aircraft's Terrain Following Radar (TFR) detects dangerously high ground ahead, it can automatically pull the aircraft up to clear the obstacle safely. The radar operates continuously during flights.

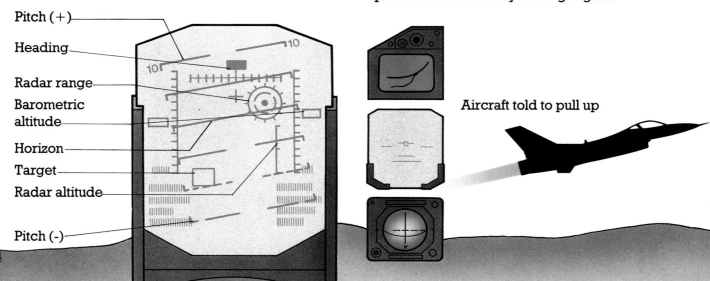

Aircraft told to pull up

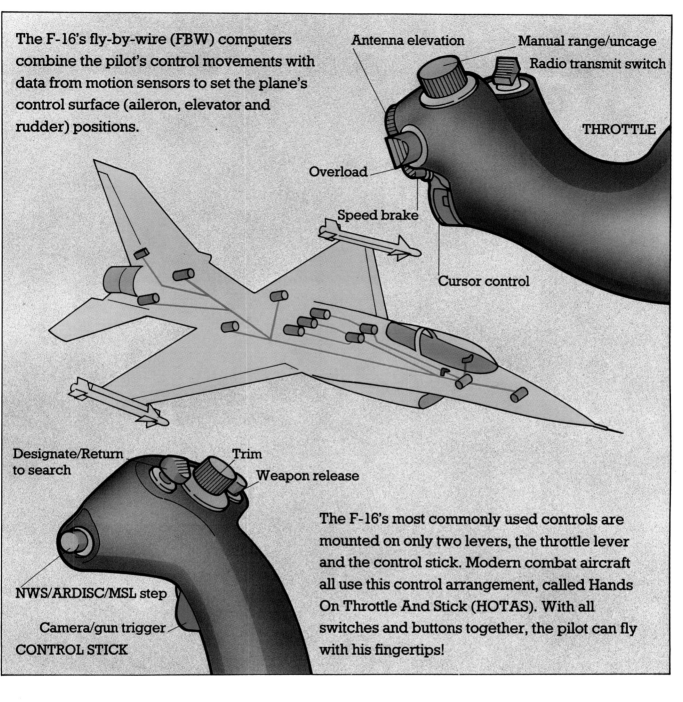

The F-16's fly-by-wire (FBW) computers combine the pilot's control movements with data from motion sensors to set the plane's control surface (aileron, elevator and rudder) positions.

Antenna elevation
Manual range/uncage
Radio transmit switch
THROTTLE
Overload
Speed brake
Cursor control

Designate/Return to search
Trim
Weapon release
NWS/ARDISC/MSL step
Camera/gun trigger
CONTROL STICK

The F-16's most commonly used controls are mounted on only two levers, the throttle lever and the control stick. Modern combat aircraft all use this control arrangement, called Hands On Throttle And Stick (HOTAS). With all switches and buttons together, the pilot can fly with his fingertips!

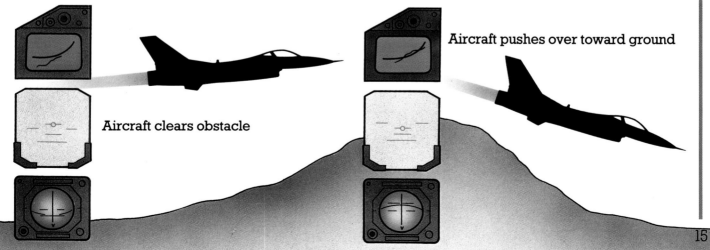

Aircraft clears obstacle

Aircraft pushes over toward ground

Electronic displays in the cockpit give the pilot vital information.

An incoming fighter uses its radar to show the position and range of any other aircraft nearby. But using the radar also gives away its presence to intercepting aircraft.

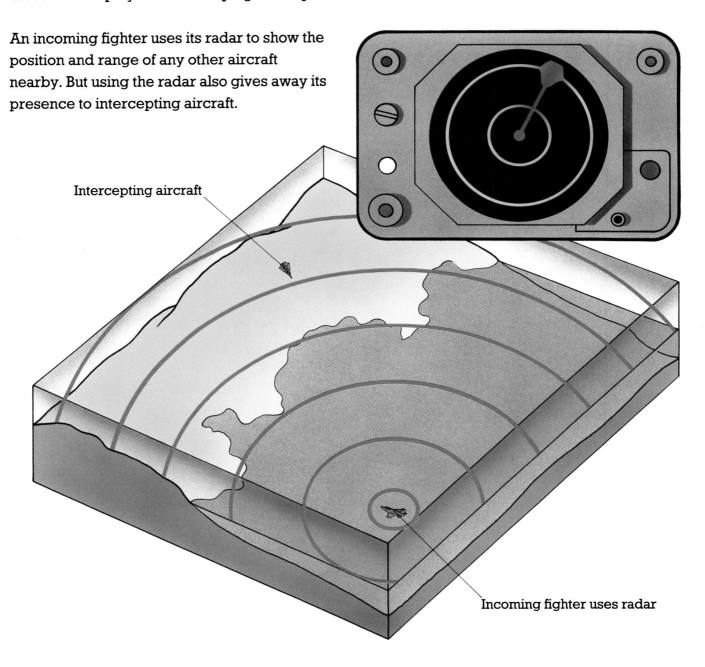

Intercepting aircraft

Incoming fighter uses radar

NAVIGATION AND RADAR

To detect the approach of enemy aircraft long before they can be seen, a combat aircraft pilot uses radar. This works by transmitting radio waves and detecting any reflections that bounce back from solid objects like aircraft. The reflections (echoes) are presented as bright dots on a television screen, showing where the objects are and their distance from the transmitter.

When a radar system based on land or on a ship at sea detects approaching enemy aircraft, messages are sent to combat pilots to fly their aircraft on a course that will intercept them. The outgoing aircraft navigate (follow the correct flightpath) using a system that senses the planes' movements, and calculates how far they have moved from their start-point and in which direction. The system can be checked and altered by signals from navigational satellites or from radio beacons in known positions on the ground.

This F-16B Fighting Falcon carries a crew of two.

WEAPONS

Fighters and attack aircraft are armed with four types of weapon – machine guns, guided missiles, unguided rockets and bombs.

A fighter's missiles are attached to the wing tips or to pods under the wings. They head toward their targets by homing in on either heat waves (infrared radiation) from the aircraft's hot exhaust or its radar reflection.

Ground-attack aircraft normally carry rockets and machine guns. The guns are fitted inside the aircraft's nose or in pods under the wings. Up to 20 rockets can be fired from a pod.

Fighter-bombers and ground-attack aircraft may also carry bombs. Some are dropped from the aircraft and explode when they hit the ground. Other bombs each contain dozens of tiny "bomblets" which are released in a pattern over the target to cover a wide area. Yet a third type of bomb is guided by laser, a powerful, narrow beam of light. The beam is pointed at the target and the bomb homes in on the reflections. "Retarded" bombs are slowed by a parachute or by tail flaps so that they fall straight down on to the target. They can be placed accurately.

Close-up of an A-10 Thunderbolt, showing its heavy machine gun.

The Matra R550 Magic Air-to-Air Missile (AAM) uses an infrared sensor in its nose to find and home in on the hottest parts of an aircraft. Heat-seeking missiles look for a particular type of radiation coming from the target aircraft. This is usually produced by the engine exhaust. Similar missiles which home in on the engine heat from ships or tanks can also be used.

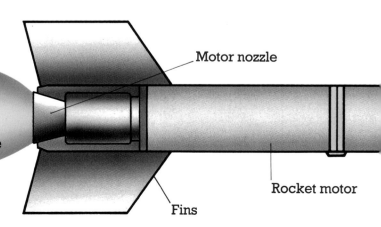

Missiles home in on their targets in one of three ways:
Active radar and infrared homing missiles follow, or track, targets without any assistance from the pilot or aircraft after launch. They are also called "fire and forget" missiles.
Semi-active radar missiles home in on radar reflections from the target produced by the attacking aircraft's radar.
Bombs may be flown to their targets by guided bomb dispensers or "buses."

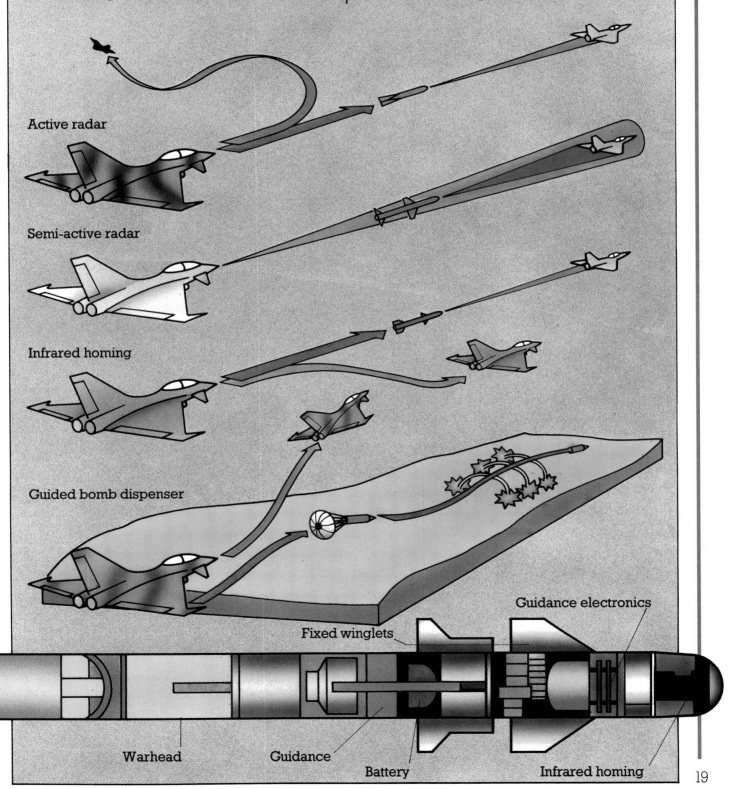

ATTACK!

Fighter pilots try whenever possible to take their airborne targets by surprise. They may do this in one of two ways.

Firstly, the attacking fighters may fly at very low level underneath enemy radar to avoid detection and only come up to attack height at the last moment. The best position for the attacking aircraft is directly behind the targets. This is because the target aircraft can only attack from the front, and their engine exhausts are in full view of the attacker's heat-seeking missiles.

Secondly, the attack force may present targets with a decoy. A pair of fighters approach the target aircraft head-on as if they are to attack. They pull away, though, without firing. While the targets are giving all their attention to the decoys, the real attack pair have flown around behind the targets.

The Thunderbolt can destroy light tanks with its gu

A surprise attack from the rear is not always possible. In the example shown, a pair of fighters have detected a pair of enemy aircraft flying directly towards them (1). The enemy craft are approaching at high speed, so the attackers act fast. The two fighters fly several miles to one side of the targets' flight line and approach the enemy craft from the side. One fighter's attack radar locks on to the first target (2). When one attacker (the "eyeball") comes within visual range of the first target, the second attacker (the "shooter") is instructed to fire a missile at the second target (3). If the missile is successful, the second target can be fought at visual range by both attackers using guns and missiles. All of this action takes place within less than a minute.

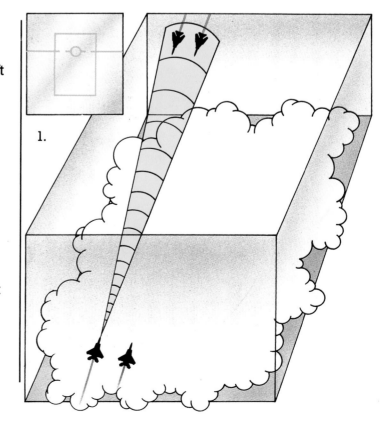

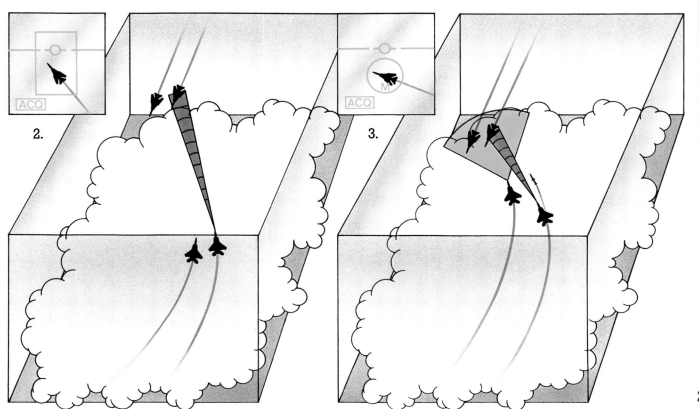

21

DEFENSE

An aircraft can be defended from a missile attack in several ways. Missiles guided by radar lock on to the largest radar reflection. A cloud of metal fibers called chaff can be blown into the air to form a larger target than the aircraft. Alternatively, a transmitter known as a jammer can send out radio signals to confuse a radar-guided missile.

Infrared homing missiles can easily be confused by the firing of burning flares out of the aircraft under attack. The missiles follow the flares instead of the aircraft.

A missile's rocket motor fires for a very short time. After that, the missile is gliding. If it has to turn in the air, it slows more rapidly than in straight flight. By twisting and turning, an aircraft may be able to outrun the slowing missile.

A Surface-to-Air Missile (SAM).

If a pilot fears an attack from heat-seeking missiles, one defense against them is to eject flares whose heat may divert the missiles.

Radar-guided missiles home in on the largest radar reflection, whether this is an aircraft or a cloud of hair-like metal fibers or chaff.

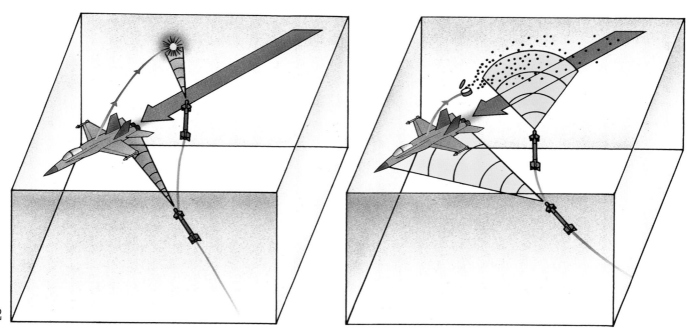

When flying in formation, fighters form a large target and are especially vulnerable.

A missile's radar guidance system can be jammed by transmitting misleading radio signals into the missile's flight path.

A missile's motor burns out soon after launch. If the target aircraft maneuvers as the missile slows down, it may be able to outrun it.

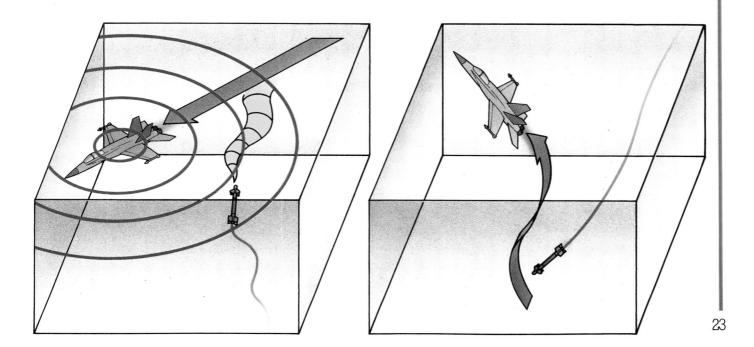

DOGFIGHT

In combat between two aircraft, the attacking pilot tries to place his aircraft in the best position to fire at the other aircraft. The defender rolls and turns his aircraft to avoid the attacker's guns. The fast moving series of attack and defense maneuvers is called a dogfight.

Every fighter pilot is trained in the maneuvers that can be flown in a dogfight to gain the advantage against an enemy. Each attack maneuver has a defensive countermove. Even with electronic aids, one pilot on his own cannot constantly watch the whole sky for enemy fighters.

Because of this, fighter aircraft normally hunt for their enemies in pairs. Two aircraft flying side by side some 2 miles apart in "combat spread formation" are better able to protect each other. Each pilot watches the other's tail, as this is the most likely direction of an attack.

If an aircraft is in danger of crashing, the pilot can save himself by using his ejector seat. First the canopy over his head is blown off. Then rockets under the seat fire and force it out of the cockpit. The pilot and the seat descend to the ground slowly and safely by parachute.

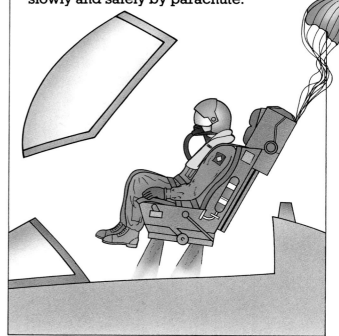

An F-16 fires a missile at high speed.

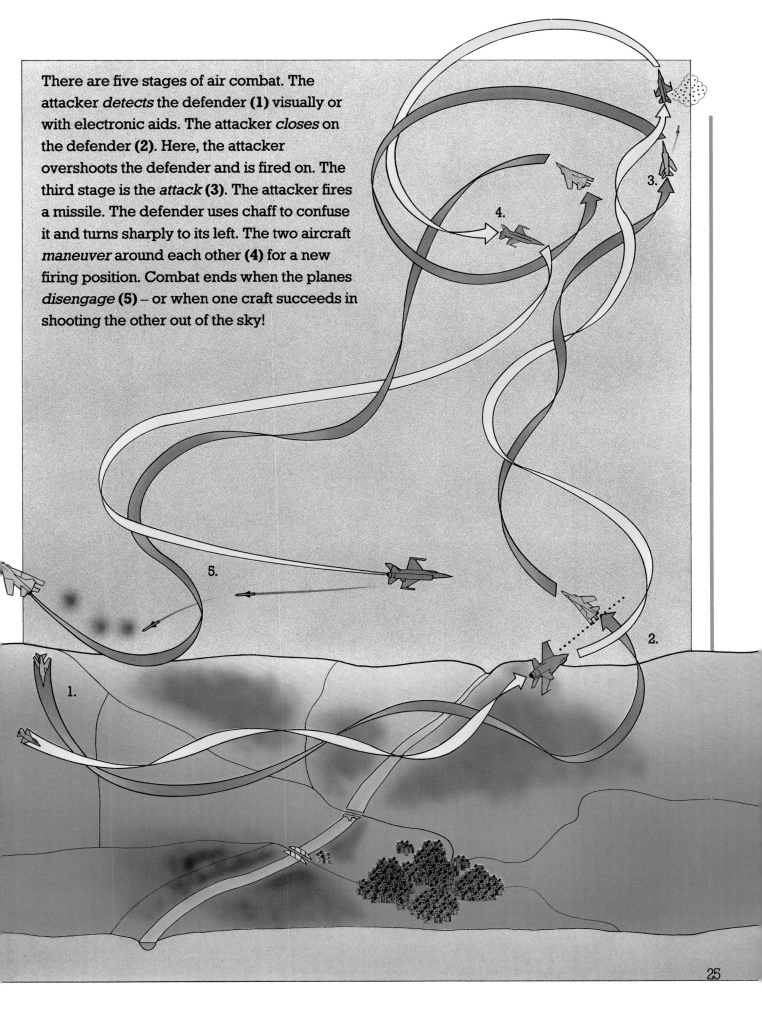

There are five stages of air combat. The attacker *detects* the defender (1) visually or with electronic aids. The attacker *closes* on the defender (2). Here, the attacker overshoots the defender and is fired on. The third stage is the *attack* (3). The attacker fires a missile. The defender uses chaff to confuse it and turns sharply to its left. The two aircraft *maneuver* around each other (4) for a new firing position. Combat ends when the planes *disengage* (5) – or when one craft succeeds in shooting the other out of the sky!

THE FUTURE

Today's research will, in the next 10 to 20 years, produce combat aircraft that look and behave quite differently from existing aircraft. They will make more use of stealth technology and electronic defense systems to avoid detection by radar. Stealth requirements will result in aircraft with a more rounded shape. The use of new construction materials will enable wings to be made thinner and to be swept far forward instead of backward. The Forward Swept Wing (FSW) fighters can be made smaller, lighter and more maneuverable than any of today's ordinary fighters.

As the number and complexity of electronic aids fitted inside aircraft increase, pilots will need more help from computers to reduce the amount of information they have to take in, think about and act upon.

Almost all existing fighters and attack aircraft depend on long concrete runways for take off and landing. The ease with which runways can be bombed may result in more Short Take Off and Vertical Landing (STOVL) combat aircraft like the Harrier.

The Grumman X-29 Forward Swept Wing fighter.

The first Forward Swept Wing aircraft, code named the X-29, was designed and built by the American company Grumman. Its computers carry out 80 control actions each second. It is a test-bed for the technology that fighters of the next century will use.

The Lockheed F-117A "stealth" fighter.

Stealth fighters have fewer flat or vertical surfaces or sharp edges than other fighters. Instead, they have smooth joints, rounded edges and angled fins to reduce the strength of the radar beam reflected back to the enemy. This makes the aircraft more difficult to detect. To date, almost all stealth technology has been developed in the United States.

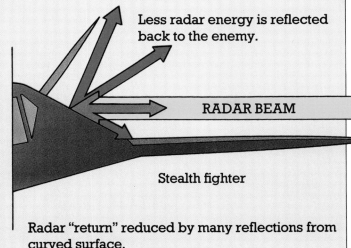

Radar "return" reduced by many reflections from curved surface.

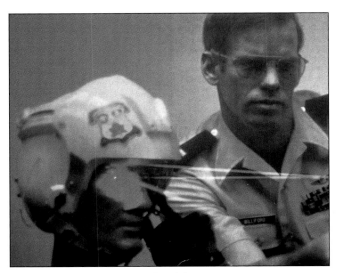

Aiming weapons by looking at a target.

Fighter-systems designers are experimenting with new ways of selecting, aiming and firing weapons. The Helmet Pointing System (HPS), for example, enables pilots to aim a gun or a missile at a target simply by looking at the target. Sensors in the helmet follow, or track, the pilot's eye movements and feed this information, via the aircraft's computers, to the weapons. The positions of several targets can be stored and tracked simultaneously by advanced versions of the system. The pilot then judges which target is the best, easiest or least dangerous to attack.

HISTORY OF COMBAT AIRCRAFT

The history of combat aircraft is almost as old as powered flight. The first controlled flight of a powered aircraft was made by the Wright brothers in 1903. Eleven years later, World War I began. Unlike today's single-wing (one per side) aircraft, many of the early fighters used in the war had two wings. They were called biplanes. Some, like the Fokker Dr. I Triplane, had three. The craft were powered by piston engines driving propellers. They were made from wooden frames covered by stretched canvas, and they carried machine guns. The propellers were linked to the gun's firing mechanism so that the guns fired only when the propeller blades were not in front of them. The pilots navigated by following landmarks such as railway lines.

The World War I Fokker Dr. I Triplane.

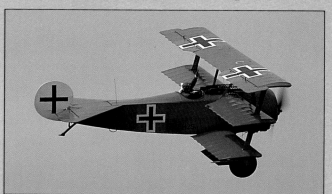

After World War I, fighter development continued. The monoplane (single wing) with a tail plane and fin at the rear became the standard aircraft design. Metal replaced wood as the material for the aircraft frame, but this remained covered by canvas until metal "skins" were introduced in the 1930s.

The Supermarine Spitfire.

World War I fighters had to hunt for the enemy by sight. In World War II, enemy aircraft could be detected by radar. Fighters were instructed to "scramble" (take off) and were guided to the enemy by radio messages. The most successful fighters of the war included the British Spitfire, the American Mustang, Germany's Focke-Wulf 190 and Japan's Zero. All were fast, agile, single-engine aircraft. In 1944, the first jet fighters, the British Gloster Meteor and the German Messerschmitt 262, entered service. By the 1950s, fighters could fly faster than sound (760 mph, or Mach 1). The F-100 Super Sabre was the first such "supersonic" fighter.

The supersonic F-100 Super Sabre.

Guided missiles enabled aircraft to attack from beyond visual range for the first time. This led to new electronic detection systems. In the late 1970s, "variable geometry," or swing wing, aircraft like the F-14 combined supersonic speeds with excellent low-speed performance. The Harrier first showed the value of Short Take Off and Vertical Landing (STOVL) fighters that do not need runways. To counter the Soviet MiG-25, the United States developed the F-15 Eagle, and the F-16 used fly-by-wire technology.

The General Dynamics F-16.

Combat aircraft are so expensive that each must be able to carry out more than one type of mission. The F-18 Hornet and the Panavia Tornado were developed as "multi-role" aircraft. In the 1980s, stealth technology has begun to influence fighter design, to minimize their radar reflections.

FACTS AND FIGURES

The gas turbine (or jet) engine was invented in 1930 by the British engineer Sir Frank Whittle.

The world's first jet-powered aircraft, the Heinkel 178, flew for the first time in Germany in 1939.

The first jet fighter to enter service was the Messerschmitt 262 in 1944.

The first and only rocket-powered fighter was the Messerschmitt 163 "Komet" (1944).

The most powerful gun carried by a combat aircraft is the Fairchild A-10 Thunderbolt's GAU-8/A Avenger. It can fire up to 4,200 rounds a minute.

The world's fastest combat aircraft is the Soviet MiG-25 interceptor, with a top speed of 2,108 mph.

A bomb was dropped from an aircraft for the first time in 1910, when the U.S. Army Signal Corps dropped a homemade bomb from an aircraft in San Francisco.

The earliest recorded ground attack was in November 1911, when Italian airmen dropped bombs on Turkish troops.

The first recorded episode of air combat took place on October 5, 1914. A French airman shot down a German Aviatik B1 aircraft with his machine gun.

GLOSSARY

Afterburning
Creating extra engine thrust by burning fuel in the engine exhaust (reheat).

Aileron
A movable flap on a wing's trailing edge, used to make an aircraft roll to one side.

Bomber
An aircraft specially designed for carrying bombs.

Control surfaces
The ailerons, elevators and rudder, the positions of which can be altered to change an aircraft's direction of flight.

Drag
A force that tries to slow an aircraft down as it moves through the air. The more streamlined an aircraft is, the less drag it creates.

ECM
Electronic CounterMeasures. The use of electronic systems to confuse or mislead enemy detectors.

Elevator
A movable flap on an aircraft's tail plane, used to make the craft climb or descend (pitch up or down).

Fighter
An aircraft specially designed for air combat with other aircraft.

Flight envelope
The limits of speed and altitude that an aircraft must stay within.

Fly-by-wire
A system in which the pilot's control movements are conveyed to the control surfaces as electrical signals processed by computers.

Head-Up Display (HUD)
A see-through glass screen in front of a pilot's eyes which shows important information about the aircraft.

HOTAS
Hands On Throttle And Stick. All the controls most often needed by a pilot are mounted on two hand controllers – the throttle and control stick.

Interceptor
An aircraft specially designed to block and attack incoming enemy aircraft.

Jamming
Confusing enemy electronic detectors by transmitting misleading radio signals.

Lift
A force that tries to raise an aircraft as it moves through the air. The amount of lift produced by an aircraft depends on its speed, weight, wing area and wing shape.

Mach
A number named after the Austrian physicist, Ernst Mach, relating an aircraft's speed to the speed of sound. Mach 2 = twice the speed of sound, which equals 1,457 mph at an altitude of 3,270 yards.

Pitch
A nose-up or nose-down motion.

Radar
RAdio Direction And Ranging. A system developed in the 1930s for detecting aircraft by reflecting radio waves off them and determining the exact direction from which they came and how long they took to arrive.

Roll
Lowering one wing tip and raising the other so that an aircraft flies forward tilted to one side.

Rudder
A movable panel in an aircraft's vertical fin used to force the tail to the left or right.

Scissors
A maneuver used in dogfighting where two aircraft cross over each other's path several times.

Thrust
The pushing force produced by a jet engine.

Variable geometry
Also known as swing-wing. An aircraft with movable wings that can be swept backward for supersonic flight or swung forward for better performance at lower airspeeds.

Yaw
A nose left or nose right movement such as a car traveling around a corner.

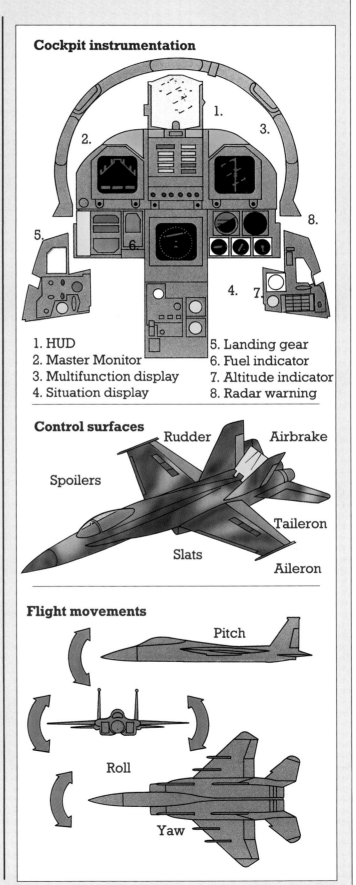

Cockpit instrumentation
1. HUD
2. Master Monitor
3. Multifunction display
4. Situation display
5. Landing gear
6. Fuel indicator
7. Altitude indicator
8. Radar warning

Control surfaces
Rudder, Airbrake, Spoilers, Taileron, Slats, Aileron

Flight movements
Pitch, Roll, Yaw

INDEX

A-10 Thunderbolt 7, 11, 29
afterburner 5, 9
afterburning 9, 30
aircraft carriers 6
air-to-air missile 18
air-superiority fighters 6, 7

biplanes 28
bombers 6, 30
bombs 18, 29

chaff 22, 23, 25
combustion chamber 8, 9
compressor 8, 9, 13
controls stick 15

decoy 20
defense 12, 24
dogfight 24

ejector seat 24
elevators 4, 15, 30
engines 4, 5, 8, 9
exhaust 8, 9, 12

F-14 29
F-15 Eagle 29
F-16 Fighting Falcon 15, 17, 29
F-18 Hornet 4, 29
F-100 Super Sabre 28
F-111 11
fighter bomber 6, 18
fighters 6, 30
flares 22
flight controls 14
fly-by-wire 14, 15, 29, 30
Focke-Wulf 190 28
Fokker Triplane 28
forward-swept-wing fighters 26
fuel 4, 9
fuselage 8

GAU-8/A Avenger 29
Gloster Meteor 28
ground-attack aircraft 7, 18
Grumman X-29 26

Harrier Jump-jet 12, 26, 29
head-up display 14, 30
Heinkel 178 29
hi hi hi mission 6
hi lo hi mission 7

interceptor 6, 30

jamming 22, 30
jet engines 8, 29

landing 11, 26
laser 18
leading edge slats 11
lift 10, 12, 30
Lockheed F-19 27
lo lo lo mission 7

Mach 31
maneuvering 6
map displays 16
Matra R550 magic air-to-air missile 18
McDonnell Douglas 4
Messerschmitt 163 "Komet" 29
Messerschmitt 262 28, 29
MiG-25 6, 29
Mirage 11
missiles 4, 29, 22
Mustang 28

navigation 17
nozzles 12, 13

Panavia Tornado 7, 29
piston engines 28
propellers 8, 28

pylons 5

radar 14, 16, 17, 22, 27, 28, 31
radar scanner 4, 5
radome 4
rate of climb 6
reheat *see* afterburning
rudders 4, 5, 15, 31

Saab Viggen 11
"scramble" 28
short take off and vertical landing 12, 26, 29
size 4
speed 6, 8, 11, 29
stalling 11
stealth technology 26, 27, 29
Super Etendard 11
Supermarine Spitfire 28
supersonic flight 11
surface-to-air missile 22
swing-wing aircraft 29, 31

tail 4, 24
tailplane 4, 5
take off 11, 12, 13, 26
throttle 15
thrust 4, 8, 9, 12, 13, 31
Thunderbolt 20
turbine 8, 9, 13
turbofan and turbojet 8, 13 *see also* engines

undercarriage 4, 5

variable geometry *see* swing-wing aircraft

weapons 6, 18, 27
weight 4
Whittle, Sir Frank 8, 29
wings 4, 5, 10, 18, 26

Photographic Credits
Cover and pages 6, 12 both, 13, 16, 17, 20/21, 22 and 23: Salamander Books; pages 7 both, 9, 14, 18, 28 all and 29: Aviation Picture Library; pages 24 and 26: Quadrant Picture Library.